Robotics and
Automation
Industry
Thought
Leaders

Robotics and Automation Industry Thought Leaders

It's been a year since the launch of the Robotics and Automation News website, and as part of our celebrations, we are moving into book publishing, the first one being this one, a collection of some of the best interviews we have conducted over the past year. It's the first of a series, as we interview interesting people from the tech world on an ongoing basis, and we thought people might like to purchase the conversations as collections contained within a book such as this one.

The original idea for the website – RoboticsAndAutomationNews. com – was to initially cover the world of industrial robots and the automation systems generally found in factories. While we have tried to keep our attention on those subjects, we also became interested in the related developments in connectivity, such as the internet of things, and computing, such as artificial intelligence.

The whole subject, if seen as an interconnected series of subject, has become quite large and involving. We can't claim to have invented anything, and we hesitate to claim any innovations in our editorial, but we do find it interesting to connect subjects together. In doing so, we find ourselves realising that subjects that previously may not have been seen as robotics-related deserved reassessment and perhaps reclassification.

For example, a car which is driven by a computer rather than a human is now generally accepted as a type of robot, whereas it might not have been before. A building which features many intelligent systems, such as responsive lighting, heating and other controls, could be classed as a robot. The building itself may not be mobile and be stationary, unlike a car, but it still has plenty of moving parts and automated and autonomous control systems.

There are other examples, but it's worth noting that there is currently a lively debate about what exactly constitutes a robot, and we try and take our lead from experts as well as add ideas of our own.

Our own view is that a robot is any hardware that has some type of software which serves as its brain. However, we recognise that this view may be too simplistic, as that would include desktop computers, which is already a classification that people recognise and accept, and we don't want to redefine things and add to the confusion.

But it's difficult to ignore what some people – including us – might call virtual robots, perhaps illustrated best by Automator, which is an application which has been included with the macOS for many years. While this type of robot is not autonomously mobile and has no moving parts, the implications of the technology are too profound and the possibilities too interesting for us not to cover it on the website.

Having said that, we try to avoid becoming another "tech" website, which usually indicates a website which deals with computing technology – desktop computers, mobile tablets, smartphones, and the software that enables them to function. There are plenty of excellent websites which deal with those subjects, and it's tough to see how any publisher could add anything significant to what's already on offer.

We try to remind ourselves that we deal with mechatronics – a combination of software and hardware which has moving parts, generally speaking. But only time will tell if we are blending things logically, or if we're making a mishmash of disparate and disconnected technologies.

In this endeavour, we appreciate the help of the many thought leaders from the robotics, automation and wider technology industry, who have been willing to spend time sharing and articulating their ideas. We believe these thoughts and ideas could, in and of themselves, lead to innovations and developments that otherwise might not have been considered and researched. We can't think of many more satisfying things than to be involved in – or at least be providing a platform for – discussions which could lead to better or more interesting technologies or ways of doing things, which in turn, may lead to better lives for everyone. ●

Chapter 2 : Jane Zavalishina

If AI Ruled the World

Interview with Jane Zavalishina, CEO, Yandex Data Factory

S ome people say artificial intelligence will eventually take over the world, and make humans obsolete and useless in large parts of society, starting with firing us from all employment – low-skilled or high. Others say it's already happened – that AI rules the world.

From artificially intelligent assistants which answer your emails and manage your diary, to artificially intelligent industrial robots that can custom manufacture products to individual customer specifications and that can tell you when they need a little maintenance, not to mention the AI in the latest cars of today which can brake and change lanes to avoid you having an accident because you fell asleep at the wheel, AI is everywhere.

And like the concept of the omnipresent deities that many religious people believe in, AI – and its attributes of machine learning and deep learning and so on – has inspired a massive global following which is growing every day.

It's growing so fast that the highly influential internet beacon O'Reilly Media, which is credited with crystallising and popularising the concept of Open Source software, says "AI is going mainstream". The company has also organised a massive event – the Artificial Intelligence Conference – to discuss all the holy scriptures relating to the new religion.

And it's not just O'Reilly. All the big names in the tech world – from Google and Apple to IBM and Microsoft – are falling over themselves to pray at the altar of the one and only true supernatural power for which we can all see and experience tangible proof.

Who wants to live forever?

AI could even give us an actual afterlife, something other religions promise but have yet to deliver in a verifiable way.

And like most other gods, AI requires something of humans. And that something is data – lots and lots of data… infinite amounts of data. For AI, as far as we know, has infinite power, because our brains, collectively if not individually, have infinite power. And it's our brains that AI is essentially emulating, albeit at a performance level that has never before even been imagined, much less achieved, by our own flawed but beloved human, all too human, brains.

In this exclusive interview, we talk to Jane Zavalishina, CEO, Yandex Data Factory, a company which counts as one of its customers Cern, the atom-smashing scientific research facility which occupies hundreds of cubic miles under the ground in mainland Europe. There was once a worry that Cern's experiments could inadvertently take humanity to hell by unwittingly creating a black hole into which we would all disappear. This has not happened. Or if it has, we've all made it through to the other side without even noticing the journey.

Facilities like Cern probably produce more data than any other type of organisation and companies like Yandex Data Factory are there to collect, collate and help scientists analyse that data. In the future it could be AI that routinely analyses the data and it could be AI that devises its own particle physics experiments. This is already happening to some extent, but is it a good idea?

Let's all panic now

AI doesn't know why it's doing what it's doing. It doesn't have humanity's best interests at heart – it doesn't have any interests at all, and of course it doesn't have a heart. Not unless these values are incorporated into it by the humans who develop the AI systems, and what guarantees does anyone have that these humans are doing that? None, that's what.

Perhaps it's too early to start worrying, but why not panic now before it's too late and AI has completely taken over our brains, automated us to oblivion and created the dystopian world so many science fiction stories have foretold?

AI will put us out of work and out of business. So why not panic?

According to Zavalishina, intellectual tasks will certainly be automated to a high degree. We started by asking about this subject.

Robotics and Automation News: **What does Yandex Data Factory do to automate intellectual tasks? What should companies look to do?**

Jane Zavalishina: Machine learning's ability to replace human decision-making by analysing data to determine the best appropriate next step for routine, repetitive tasks means that it is ideally placed to lead the next generation of automation. Using data to determine the best course of action within a set of finite rules is one thing, but empowering technology to draw its own conclusions, by learning from empirical data – is quite another. This is where machine learning offers a competitive advantage through decision-making and automating intellectual tasks by providing greater precision compared to rule-based systems and human analytics.

To give an example – let us look at maintenance efforts that are performed for any equipment or machinery. Typically, the decision to execute maintenance or replace spare parts is guided by a pre-defined schedule and a set of conditional rules. Such rules are often guided by previous statistics of faults or breakdowns, but they are widely generic and tend to be suboptimal in terms of costs incurred. In fact, maintenance is always a balancing act, as both excessive support efforts and potential breakdowns and delays mean potential losses.

Equipment maintenance is a good example of a process that can be automated with the help of machine learning. Instead of simply following a predefined schedule or responding to alerts, a machine learning-based predictive model can be built. This complex model can precisely predict potential breakdowns before they happen or before they are detected through human analysis. This is achieved through the analysis of past maintenance data, performance and machine telemetry. This also allows for the ordering of spare parts to be done automatically. Eventually, this reduces costs and automatically streamlines servicing efforts.

When it comes to machine learning, it's important to remember that the more relevant the data fed into decision-making, the more accurate and appropriate the decision made. But equally, the bigger the amount of data, the greater the need for machine learning and automation due to the sheer complexity of the calculations and impossibility to properly leverage the data without the help of machines. That is what we offer at Yandex Data Factory – the development of such machine learning models that help companies automate routine decision-making to achieve measurable business results.

Companies are investing in machine learning not because it's a fad or because it makes them seem pioneering. They invest in it because they are seeing the opportunity for rapid positive return on investment. Being able to predict success more accurately or avoid potential losses preemptively are game changing benefits and something companies will always strive for.

What in the long run do you think will be a challenge for AI developers? What is the intellectual quality that humans have that robotics/automation/AI will not now or possibly ever be able to emulate?

The greatest challenge for AI is for it to explain its actions – why it has done what it has. While machine learning is capable of automating many intellectual tasks, AI developers are faced with the challenge of replicating the human brain and rationale. No small feat.

The main challenge lies in establishing trust towards AI – as it is often impossible to rationalise machine learning in the same manner we rationalise and justify our actions – simply due to complexity of decision-making and a vast number of factors taken into account. When it is impossible to understand the "thought process" behind the machines, the only way to build the trust is through experiments demonstrating and measuring the direct value brought by the smart algorithms. Establishing this experimental culture in business practice will lead to greater trust and, as a result, a greater division of labour, and increasing readiness to accept machines without fear, doubt or scepticism from humans.

However, there is one thing machines still cannot do: creativity and defining the best actions in the absence of past historical data. This strategy guidance and bold risk-taking is still reserved for humans, and will continue to be for a while yet.

What are the economic implications for AI – good or bad?

Data compiled by Deloitte from the census data for England and Wales, stretching back to 1871, has shown that technology has created more jobs than job losses. Nonetheless, some have come to fear AI due to a belief that as automation becomes more sophisticated, the need for human input will decrease and jobs will be lost.

In fact, while robots may replace some roles, this shift to relying on technology will enable people to focus more on creative and strategic tasks, changing the way they approach various processes and even changing industries altogether. The first step to this is for humans to assess which tasks machines will take on and which strategic and creative tasks are to be dedicated exclusively to humans.

This shift in the way humans and organisations work together with machines will ultimately see the economy boosted by increased efficiency. As tasks become automated and efficiency increases, we'll see cost reductions on a large scale, improving entire industries and propelling the economy forward.

On the bright side, automation is likely to improve the quality of life, especially for low-income households. Many things will become more affordable, and personalisation, facilitated by AI across various industries, will become a commodity.

Will humans ever become slaves to the machines? Could the growth of AI have a negative impact on society?

AI and automation will continue to evolve and its sophistication will heighten but far from becoming slaves to machines, humans will be freed by them. With the onset of AI and automation, we will see machines take over the mundane which will leave humans with more time and energy to explore different goals.

Moreover, as costs are reduced, businesses will have more money to spend on things which may have otherwise been compromised in favour of profitability – for example, corporate social responsibility or ethical activities.

If humans want to make the most of AI, we must ensure that we learn how to use these new technologies to their full potential so that they support our transition into strategic beings. ■

Chapter 3: Minoru Usui

We Are Living in 'Extremely Transformative' Times

Interview with Minoru Usui, president, Seiko Epson

When you want to talk to someone about robotics and automation, especially in an industrial context, who better to talk to than an expert from Japan? Not wishing to compound any stereotypes, the "land of the rising sun" has shown itself to be an early adopter of many, possibly all, advanced technologies for many decades now.

When George Devol and Joseph Engleberger invented the first industrial robotic arm in around 1950, Japan was the most enthusiastic buyer of their product, the Unimate. Devol and Engleberger had found business tough in the US, where there was a general perception – perpetuated by Hollywood science fiction films – of robots as being an otherworldly, threatening menace. Which they are, of course.

It's inevitable that robots will take over the world and more or less enslave us, but when have humans ever listened to warnings of our impending doom?

There was a man who used to walk along Oxford Street, in London for many years during the 1990s, with big white message boards strapped to his front and back. The message he carried in big black letters was, "The end of the world is nigh". Did anyone pay any attention? Of course not. Proof, if it were needed, that we just don't listen to people who can see the future and make it clear to us in black and white, until it's too late.

Maybe he got the timing wrong, but that guy was onto something.

Our world is ending, and the robot world is beginning. Should we be worried? Of course we should. But what can anyone do to stop the march of progress? Mass unemployment caused by increasing use of robotics and automation technologies, and let's not even get into the frankly apocalyptic prospect of ubiquitous artificial intelligence running, ruling and possibly ruining our lives. What can anyone do against the overwhelming forces of human nature, the drive for progress, the greed, the lust for power and money, especially when it's combined with such fascinating and alluring, not to mention lucrative, technology?

But maybe we shouldn't worry too much. Maybe the robots come in peace. Maybe they just want to help us become more productive, give us more free time to enjoy our lives, sometimes known as unemployment. Maybe the change they will inevitably bring will be a gradual process, providing us with plenty of opportunity to adjust and adapt. And maybe we don't have to study where the Luddites went wrong so we can do what they did, only better.

Whatever we decide, the societal upheavals heralded by the accelerated development of AI-based technologies are not only inevitable, they are already happening.

In this exclusive interview, Minoru Usui, president, Seiko Epson, gives Robotics and Automation News his views of the "transformative" effects of robotics on the world today, and looks forward to what dreams or nightmares may come.

Epson is one of the world's largest robotics companies. The

company has more than 50,000 robots installed around the world at last count, which was at least several months ago. Given that the past year has seen a huge hike in demand for industrial robots around the world, especially from China, Epson probably has much more than 50,000 robots installed globally by now.

In the UK, the company sells many SCARA and 6-axis robots. While Britain is fond of self-deprecation and much is made in the media of what some observers – mostly "loony lefties" – see as the deliberate decline of the manufacturing industry over the past few decades in favour of becoming a more service-oriented economy, mainly based on financial services, the fact is that the country is still one of the top 10 manufacturing nations in the world, with annual output valued at around $245 billion, according to government figures.

However, while it may be a top 10 manufacturing nation, the country is somewhere near number 20 in the global list of robot density, and this is one of the main reasons for the relatively low levels of productivity in UK manufacturing, according to the government's Office of National Statistics.

From Minoru Usui's point of view, the UK, much like other advanced economies, can gain huge benefits from adopting more robotics into the manufacturing sector. The world is changing, he said, and Britain could change along with it. Usui began his career at Epson in the late 1970s, which is about the time historians would probably say UK manufacturing jobs moved to lower-wage, developing nations, and large parts of Britain's manufacturing industry started shutting down, throwing millions of workers into unemployment and the country into political and social chaos, culminating in the Winter of Discontent. But there is a chance that things can improve now, according to Usui.

"We're living through an extremely transformative period, characterised by major geopolitical, economic and technological changes," says Usui. "Shaped by these mega-trends, value chains are being redistributed as production centres move closer to market; often back toward developed nations as the global wage gap continues to shrink. The shift of manufacturing back to the US and European markets is well under way and robots can play a huge role in this re-shoring trend."

One crucial factor in this possible reversal of manufacturing fortunes and employment prospects, says Usui, is the fact that robots are approximately the same price for everybody.

"Their cost being the same wherever they are used, robots can make a big difference in the spread of total labour costs in the UK and developed markets. In addition, robots are getting cheaper, more dexterous and easier to operate and could therefore help tackle the looming issue of labour shortage in the manufacturing industry. Above all they are capable of delivering improved productivity and can reduce the need for outsourcing."

One perennial concern of British people is the cold weather. In the 1970s and even through the 80s, most people did not have central heating, often using coal to heat their homes – not because of post-modern aesthetics but because their toes were freezing. Bitterly cold winter weather made some jobs difficult, dangerous or even impossible. Even today, if there is any news that there may be a fuel shortage, something close to panic ensues and the government is placed under pressure to avert whatever catastrophe might occur.

Usui says the UK can get by with a little help from robots. They could play a critical role in ensuring that British people's toes, and the toes of everyone around the world, are kept warm. And they have the potential to be even more helpful in other ways and in other industries.

He says: "In addition to manufacturing, robots are likely to play an important part in tackling global trends such as securities shortages – in particular fuel – and the ageing population.

"Autonomous robots come with the potential to handle a great number of tasks that require a precise, delicate touch, from food handling to managing dangerous or volatile substances. Longer term, robots will be suitable for deployment in high precision or high risk environments where humans cannot, or should not, operate.

"Power generation is a notoriously risky business – from mining to fracking to drilling for oil, to leveraging the opportunity of nuclear solutions – these industries have always had their risks. What if autonomous robots could take on the most dangerous of these tasks and be used to work in the most high risk locations – a nuclear reactor, for example? That possibility is approaching faster than you might think."

Mining is no longer a large-scale activity in the UK, after Margaret Thatcher swung her formidable handbag and dealt the industry its final, decisive blow during her first term as prime minister, starting in 1979. The miners went on strike and the clash went on for around a decade in what was one of the defining political confrontations in the country's modern history – lefty unions on one side, right-wing bosses on the other, Thatcher being one of the most right-wing leaders this country's ever had.

To some it may seem the recent past in terms of time – it was only the 80s. But when you look at how the economy has changed, it's difficult to imagine the coal industry being such a large and important employer and industry in the UK. These days, the talk is all about "digital Britain", tech this and tech that. Silicon Square, Tech Triangle, Robot Roundabout. It's a completely different place.

And it's likely to change even more. The robots have only just now started to emerge in large numbers, ready to take over the world, having been isolated in factories for so long. They also have new connectivity technology, a robot religion if you will, called internet of things (IoT), which allows more than three robots to congregate at one time, something Thatcher may not have allowed.

Plus, Britain's population, just like populations in Japan, Germany, the US and practically all advanced economies, is ageing. Around 10 million people in the UK are over 65 years old, according to government figures. That's one in every six people. And the government is having to think about how to look after them long term, particularly if large corporations increasingly avoid paying taxes. The National Health Service might be free to end users, but it's not cheap to maintain from the point of view of those who do pay their taxes.

Usui believes robots could be the revelation the world has been looking for. "My future vision sees the capabilities of robotics extending into areas like assistive care, with robots helping us to address the challenges of supporting an ageing population while reducing the cost implications – particularly when you consider their potential once connected to the internet of things.

"As robots, along with other technology products, become increasingly connected, collecting data from all around them and relaying it to the internet – as well as accessing and learning from data gathered by other machines – they will become a very real solution to supporting the service and care industries. This support will manifest in multiple ways, not only with basic manual tasks such as lifting and moving, but also more complex tasks such as measuring and sensing for environmental data, recording and reporting changes and fluctuations, and providing data based recommendations and initiating tasks to apply them.

"As they progress further to add artificial intelligence to their capacity to learn, they will be a key solution in providing services within the home and care facilities, alleviating resource and labour pressures and helping to improve service delivery and patient care."

Clearly Usui does not think the robots will use the IoT to organise a revolution, or a global takeover. He doesn't even think they're clever enough.

"My belief is that robots will never be truly 'intelligent' – only ever artificially so – and will always rely on humans to ultimately programme and 'drive' them. Robots, regardless of how clever, will remain tools. That said, they are tools with enormous potential and power, and we must respect any technological advancement that has the potential to alter social dynamics. It is our responsibility to ensure that they drive positive change.

"I have no doubt that robots are destined to help us solve major societal issues such as labour and skills shortages and costs, and will finally realise their potential to release people from ordinary, dull or dangerous manual work, freeing more of us to concentrate on work that demands human intelligence. I have just as little doubt that they will be one of the keys to us solving numerous other fundamental global issues, relieving the pressure on our more finite resources."

Usui has been working with robots for almost 40 years, so he should know what he's talking about. He is obviously optimistic about the future, and probably imagines a world where man and machine will live as one.

But then, that guy with the sandwich board also knew what he was talking about, and I still say he was onto something. ●

Chapter 4: Saagar Govil

Electrifying New Markets

Interview with Saagar Govil, chairman and CEO, Cemtrex

Alot's happened since this website was introduced to Cemtrex a few weeks ago. The company's been on a gigantic spending spree, buying up companies and restructuring its business for new markets as though it were in a hurry to get somewhere fast.

Where that somewhere is may be deduced from the acquisitions Cemtrex has made and the types of products and services in which the acquired companies specialise.

Last month, Cemtrex purchased an obscure German company called The Target, an electronics manufacturer which supplies top-level automakers.

Then, earlier this month, Cemtrex bought up and is synergising its operations with another German electronics manufacturer, Periscope. Not to be confused with Periscope the video streaming app, the Periscope Cemtrex bought is another supplier to major automotive companies.

It doesn't take a genius to figure out what Cemtrex has its heart set on – the market for electronics used in autonomous vehicles.

However, in this exclusive interview, Saagar Govil, chairman and CEO of Cemtrex, indicates that the company's ambitions are not limited to driverless cars and such – they include other nascent and exciting markets in what might be called "smart" clothes, except "smart" in this context means they're loaded with electronics which enable the clothes to gather information and maybe even change its state.

So, for example, a jacket worn by a person taking a walk outdoors may have electronics which heat up the jacket if the weather is cold and cool down the jacket if the weather is hot. The safety issues surrounding such wearable technology is certainly worth looking into, especially as the idea seems good and popular in principle, and these "smart" clothes are likely to be worn by millions of people very soon.

Alpha geek

Cemtrex, which also owns brands such as ROB and Griffin Filters, hasn't yet made all the financial details of the Target and Periscope deals public, but SeekingAlpha.com calls them "major acquisitions", adding that "Cemtrex management continues to buy when there is blood on the streets", possibly referring to market conditions in manufacturing not being as positive as they might be.

Cemtrex is listed on Nasdaq, and has a market capitalisation of around $30 million. Its share price saw a healthy increase following the recent corporate acquisitions, which seem to be considered strategically sound by investors.

And the company isn't just buying big, it's talking big too, publishing its own comment on India's decision to ratify the Paris Climate Accord, which is some sort of agreement to reduce "greenhouse gases" or "global warming" or something – it's probably what was known simply as "pollution" back at the beginning of the millennium when Cemtrex was established as a subsidiary of Ducon, one of the largest Indian-owned companies in the US, with annual sales of more than $400 million.

As an independent concern, Cemtrex made a profit of approximately

$3 million in 2015. Its annual report says the company's net income has grown more than 20 per cent a year over the past five years. Around 300 people are employed there, a lot of them engineers.

Our type of company

Robotics and Automation News isn't big on financial stories, but some developments are worth noting, especially if engineering innovation results in the growth of a company's market value. And Cemtrex looks like it's on a growth trajectory. So we asked Govil to provide us with an overview.

Robotics and Automation News: Which products and services have helped Cemtrex achieve business success over the past year or more? Which ones have performed well, and why?

Saagar Govil: "Electronics manufacturing services have been a large part of our recent growth and we expect it to continue for the long term. We are the supplier of electronics of some of global leading European OEMs [original equipment manufacturers], particularly two that are leaders in robotics and automation.

"Additionally technological advances from connectivity and data processing have enabled us to implement new environmental and process monitoring technologies that operate in real time rather requiring laboratory analysis in the past.

"This is a real game-changer for many industrial plan operators looking to improve efficiency and process optimization.

"We focus our efforts into this rapidly growing markets in order to achieve our long term growth objectives. It is a strategic approach to position our company into these markets in which we see growth due to innovation occurring and we can subsequently add value."

If you're going to hell, at least wear a cool jacket

Cemtrex supplies printed circuit board assemblies, provides instruments for industrial processes, and provides industrial environmental control systems. Given that almost no complex, engineered object in today's world is purely mechanical any more – every one has a chip in them or incorporates some electronics, except for some very expensive Swiss watches – the market for companies like Cemtrex would seem vast.

Of course there's a lot of competition – there are many electronics companies out there. But new markets – the one created by the emergence and expansion of the internet of things, for example – are providing opportunities for many of these companies to grow.

Robotics and Automation News: How will your products change to accommodate connectivity of machines – industrial internet of things? What are your own observations about IIoT?
Saagar Govil: "Companies are able to implement technology into things that never had them before. For instance, one of our newest customers is an Italian leather apparel manufacturer particular for motorcycle racing markets.

"They came up with the idea to implement electronics into the jacket such that when a motorcycle driver is at risk of toppling off the motorcycle, an airbag in the jacket would deploy, attempting to save the driver before he or she hit the ground.

"This is an extremely innovative piece of wearable technology that is only possible due to recent technology advances and is also an example of how companies that previously had no business in being a tech company suddenly have a whole new set of products in their business.

"We are seeing this occur in the IoT market as well with appliance manufacturers building in connectivity so that these products can provide data and communicate with the cloud. The motorcycle jacket also has connectivity both with smartphones and the motorcycle itself and there are a number of applications that this device could be tailored to, like elderly people at risk of injury from falling down.

"With environmental monitoring, industrial plants are allowing our analyzers to communicate directly with the cloud so we have remote access to both service the unit as well as provide them data. Some of these refineries were built 100 years ago and are operating on technology decades old. This new level of sophistication provides them with tremendous savings and efficiencies in their production processes which will ultimately lead to high profitability for them.

"We are seeing a direct ROI [return on investment] for these types of applications. Better connectivity in all types of industrial devices is a growing trend – we are seeing a lot of this from our industrial OEMs. We see recent IoT products as just the beginning of a whole new breadth of technological integration into otherwise everyday consumer products as well."

The difference between US and Germany

Being an industrial company, Cemtrex has a good global perspective, so we asked Govil to comment on how leading industrial nations differ – perhaps in the way they do business, or their attitudes to technology.

Robotics and Automation News: What has been your experience of the differences between doing business in different countries?

Saagar Govil: "I can comment on my experience on the US and Germany. Germany retains a culture focused on innovation and implementing it, which I find you don't see here in the US outside of Silicon Valley. For instance, it is expected that Germany will allow driverless cars before the US does.

"However, the advancements that the US puts out through Silicon Valley focuses heavily on software and its disruptive impact into a variety of industries, and the US is the global leader in that respect. Typically innovation in the US is driven by startups, which are subsequently acquired by larger companies.

"In Germany you find the innovation is driven directly by the major companies, albeit at a slower pace, in my opinion, because these are such large companies.

"I think this difference between the US and Germany is a result of the business risk. In the US, its easier to be an entrepreneur and bootstrap your way up. In Germany, due to labor laws and other factors it's less easy to maneuver the challenges startups face."

What do we know?

Cemtrex deals with some of the largest companies in the robotics and automation industry, although we hadn't heard of it until recently. Part of the reason for this is journalistic procrastination on our part, but also partly because Cemtrex doesn't like to publicise who its clients are, so we didn't connect them to our audience.

They've mentioned some names to us off the record which made us realise that editorial coverage of Cemtrex would be consistent with our editorial policy, if only we had one. We're working on it, but you know what startups are like – too many things to do too fast in too many directions. But we'll get there/bought/wound down in the end. ■

Chapter 5: Arun Srinivasan

'Bish, Bash, Bosch'

Interview with Arun Srinivasan, senior vice-president, Bosch

It's a small word. Bosch. Associated with so many things, most famously "bish bash bosch". A meaningless phrase really, but according to the Urban Dictionary, it's "used to describe the efficiency of a process you have just explained, often used if there are three steps to the process".

For example, someone might say: "So there you have it. Clutch down, first gear, handbrake off and you'e away. Simple as that, bish bash bosh."

If you want to be pedantic, the correct colloquialism is actually "bish bash bosh". We made the other one up, although we only added one letter – c. We thought it would make a more interesting way open a feature about a company millions of people are already familiar with as being one of the largest industrial companies in the world.

Bosch's revenue for last year was more than €70 billion, and the company was founded in the late 1800s by Robert Bosch. It's still more than 90 per cent owned by the family trust.

It's all about the letters

These days, to some people, €70 billion may not seem like a lot of money, what with so many dotcoms being valued at trillions. But Bosch is one of the 30 largest manufacturing companies in the world. Remember manufacturing? It's where you make stuff. On a large scale. It's becoming an increasingly rare activity in some countries that are considered "advanced economies", manufacturing having at some point become a concept associated with developing countries which pay low or next-to-nothing wages.

Manufacturing tends to be thought of as labour-intensive, which means high costs. Hence the rise of China, with its 1.7 billion population, and other highly populated nations of the East. Meanwhile, the West has seen jobs and industry drain away, leaving vast areas of some countries decimated and entire ways of life and cultures diminished if not destroyed.

But things could be coming full circle. Manufacturing jobs may return to the West. Sounds strange, but dreams do sometimes come true. It's not necessarily a flight of fancy. Some very knowledgable people believe in the possibility. Opinion is divided on the issue it must be said, but it's better than being unanimous against.

Those who believe there may indeed be at least steady growth in the manufacturing industry in Europe and America point to robotics and automation technology as the possible saviour.

Barclays Bank conducted a survey of the manufacturing industry in the UK and found that an additional £1 billion of investment into robotics would lead to a £60 billion increase in the nation's economic output.

It might have been around a long time, but robotics and automation technology is only now becoming accessible to smaller business operations – the systems are getting cheaper and more capable of doing more things.

Moreover, advances in computing mean that a relatively small company could, for example, install robots around the world, connect them to some sort of cloud service using the internet of things and manage them to make whatever products it's geared up for.

What's in the name?

Take the Liam robot, recently unveiled by Apple. The 29-arm system can completely take apart an iPhone, right down to its constituent components and materials, in 11 seconds, according to the company. That's eleven seconds. And it's not like a smash-and-grab raid. Liam can save all of the components and materials for recycling, which is exactly what Apple designed it to do.

But imagine a robot called Mail, exactly like Liam, but in reverse. Mail is a robot that puts all the constituent parts and materials of a phone together. That is, if you give our Mail robot all the bits and pieces needed to make an iPhone, it could potentially make a smartphone for you before you can make a coffee.

So what's the message here? Maybe you'd have to ask Apple for a fuller answer. But some implications are, needless to say, a lot of unemployed people in Asia, and potentially many more startup companies offering jobs for roboticists and other workers with relevant skills. The only other thing these companies will need to be globally successful is to build an image and product line as popular as that of Apple, which appears to have given a new meaning to reverse engineering.

Reverse-engineering the future

You can't reverse-engineer something that doesn't yet exist. Maybe you can. But the people we know can't. Although, having said that, maybe engineers and inventors and creative people in many fields are doing exactly that – they are reverse-engineering objects and ideas of the future as they imagine them.

It would be interesting to know what the first examples of autonomous vehicles were. ComputerHistory.org seems to suggest that the flying carpet was the first autonomous vehicle. It might have some sort of mystical power on its side, but autonomous flying carpet technology was never really going to catch on – no seats, that's the problem.

Whatever the origin of the species, the driverless car is combusting itself through science fiction writers' highly imaginative mindscapes right down onto the very real roads of the cityscapes in today's world. And many companies are betting big on the successful proliferation of the technology.

With more chips to play than most, Bosch is very much at the table, and seems to be playing its cards with the adroitness you might expect of a startup company. But then, everyone's still at the drawing board when it comes to driverless cars.

Long in the making business

The vast amount of hype notwithstanding, autonomous vehicles are still not ready to take to the roads. But if any company has the resources and infrastructure to provide crucial technologies to bring autonomous cars into the real world, it would have to be Bosch.

The company has expertise in various markets, such as power tools, household appliances, security systems and thermotechnology; it makes drives and controls for the robotics and automation systems that themselves manufacture or package other goods; but perhaps most importantly, Bosch is said to be the world's largest supplier of automotive components.

Its history includes innovative developments of the earliest forms of combustion engine, back in the early 1900s. The company itself was established in Germany in 1886, and today employs close to 400,000 people all around the world.

But far from allowing its illustrious history to lull the company into complacency, Bosch appears to be working hard to establish itself as a leader in this newfangled autonomous vehicle technology business. A number of new initiatives in the past few months would suggest a mixture of enthusiasm and perhaps a sense of anxiety on the part of Bosch. For no matter how big and venerable a company is, the world of technology moves so fast and so dramatically that no one can really stay ahead of the game unless they make it central to their plans to do so, and even then there's no guarantee.

Bosch, Earth

The idea of AI- or computer-controlled, internet-connected and fully-autonomous cars has placed Silicon Valley tech companies in the driving seat, with the world's traditional automakers playing catch-up.

The prestigious car marques of Germany, with more than a century of history behind them, are all exploring this newly forming vehicle sector and thinking about what they need to do in order to maintain their power into the next few crucial decades.

Bosch itself seems to have placed driverless cars front and centre of its plans going forward. It's already Planet Earth's auto parts supplier, and even present-day human-driven cars are increasingly becoming autonomous. But its empire was built on combustion engines and human-driven cars. How does it plan to evolve into one that may find that the electric engine and AI-driven robots are the biggest part of its business within the next 20 or 30 years?

The man who knows enough

In this exclusive interview, Arun Srinivasan, senior vice-president at Bosch, talks about the "big business" of car technology. We started by asking how Bosch is positioned in the autonomous vehicle technology market, how important this market segment is to Bosch considering how diverse the company is, and what competitive pressures it may be facing going forward. How, for example, can Bosch compete with tech companies with their background in computing, digital cartography, and artificial intelligence?

Srinivasan says: "Back in 2011, Bosch set up a team that has been working exclusively in the area of highly automated driving. Bosch prototypes are currently testing on public roads in Germany, the US and in Japan. We plan to make highly automated driving available in 2020. Driver assistance, which is the foundation of automated driving, is already big business for Bosch. We expect sales of €1 billion in 2016, just in that area."

Stop in the name of AI

Many cars on the road today already feature a significant amount of autonomous driving technology, the best-known and perhaps most appreciated of which is autonomous emergency braking, fast becoming a standard part of the AI and electronics systems in many new models produced by leading automakers.

Srinivasan says: "Over the past 20 years, Bosch has been one of the drivers in bringing electronics and software into the car. As the car is fast becoming part of the internet, automotive technology is merging more and more with telecommunications and consumer electronics.

"Bosch is confident of making its mark in this area. We have extensive systems expertise relating to the car. We offer technology that connects the car to the internet, and are developing various service solutions around the car. For example, connected parking and a cloud-based wrong-way driver warning."

Predictive catalytics

Fully autonomous, or self-driving, cars may be 10 or more years away. It could happen sooner if lawmakers can devise a regulatory infrastructure that protects the general public from what would be an army of autonomous robots on the road.

One of the first fundamental changes in automotive technology since the invention of the combustion engine happened in the 1990s, when legislators required automakers to fit their vehicles with catalytic converters to reduce the amount of dangerous fumes they release into the atmosphere.

The move came after decades of growing concerns about pollution. Now, having moved from the margins to the mainstream, environmental issues will probably lead to electric cars becoming the most common form of motoring. But there's no such life-saving imperative behind the interest in autonomous vehicles, although there are those who claim that AI systems would be safer drivers than humans.

Are we there yet?

But what is Srinivasan's best guess as to when autonomous cars will become commonplace on the roads? He should have a better idea than most, so we asked him. We also asked him what the main hurdles are – regulations, cybersecurity, or something like that? In the meantime, while we wait for fully autonomous buses to arrive in batches of four or five, how does he see vehicle automation within the current human-driven cars developing, and how is Bosch doing in this market?

"In order for new features to be accepted by drivers, they must spark enthusiasm whilst being safe and easy to use," says Srinivasan. "The behaviour of the features should be understood just as intuitively as their limits.

"Bosch currently focuses on highly automated driving on motorways. Full autonomous driving – also in inner-city traffic – is clearly more than

10 years away from now. The trend towards automated driving poses new technical challenges with regard to the surround sensor concept, steering and braking, and the electrical or electronic architecture in the vehicle. The communication protocols for data exchange between vehicles must also be standardised.

"However, the greatest challenge in the development of autonomous vehicle functions is still traffic in urbanised areas, where an extremely wide range of road users and obstacles must be taken into consideration, and all around the vehicle. The respective algorithms must work reliably, carefully and consistently in all situations.

"In addition to the technical challenges, the legal framework must also be clarified in order to pave the way for automated driving. We are confident this will happen, as the necessary legal changes have now been initiated within the European Union. Finally, cybersecurity is a prerequisite. Bosch has technical solutions that offer a high level of safety and security."

Die hard sell

Bosch doesn't need to sell itself much in some markets – it more or less dominates them. But autonomous car technology is an area where no company has a significant history, so in a way it's all to play for. Perhaps this is why Bosch is increasing its prowess in this area.

But we asked Srinivasan, why autonomous cars? It's an unfair question in a way, considering that so many companies are going into that business, and not many would question their motives. But Bosch has such disparate interests, how can it avoid being seen as not specialist enough? How does a company like Bosch build confidence in customers for a technology that Bosch – and everyone else – has no history in?

"Although it has advanced significantly, autonomous driving actually started with active safety systems and driver assistance," he says. "Bosch clearly paved the way for the industry: Bosch was the first company to introduce an electronically controlled ABS – the Anti-lock Braking System – in 1978; it invented the Electronic Stability Control in 1995; and the radar-based Adaptive Cruise Control in 2000.

"Further safety and assisted driving functions followed and we are now leading in some of the connected technology too.

"Bosch has unique expertise in vehicle and systems technologies covering Powertrain, Braking, Steering and surround sensing. We invest heavily in R&D, €6.3 billion in 2015. Collaborating with a number of other companies, Bosch will research, develop and test new functionalities for cars and trucks, offering both partially automated and highly automated driving on motorways, in urban scenarios, and for close-distance manoeuvres."

Bosch makes billions from autonomous tech

Bosch reported sales of €70 billion for the first time last year – a record. And part of the reason for the growth in revenues is that Bosch is actually making money from autonomous driving technology well before fully autonomous cars are on the roads – just partial autonomy will do for Bosch thank you very much.

We asked Srinivasan to give us some forecasts about how much Bosch expects to make in automated driving, or autonomous car, technologies going forward.

"Over 60 per cent of our turnover is in the area of mobility solutions and sales of automated and driver assistance technologies have been growing steadily over the past few years," he says. "As the industry moves further towards automated driving, the market for driver assistance systems is expanding with the technology now becoming available in all vehicle classes.

"An important driver also is the Euro NCAP rating, which focuses increasingly on technology to make driving safe for everybody – for example, driver assistance systems such as the automated emergency braking. This is now a pre-requisite for passenger cars to receive a five star rating.

"Bosch is increasing its sales by a third each year, in this market. Last year was the first time that the company sold more than 50 million environment sensors for driver assistance systems. The number of radar and video sensors sold has doubled in 2015 – as it did the year before."

Towards a safer, more automated world

Srinivasan is on the Bosch board in the UK, where he is based. Recently he steered the company into an autonomous vehicle technology collaboration project which will see smart mobility being tested on the roads and pavements of London.

The project is backed by the UK government, through an agency called InnovateUK, which is investing tens of millions in the future of autonomous vehicles, hoping to keep up with the rest of Europe and indeed the world.

Srinivasan is an enthusiastic part of the project, partly because he believes autonomous driving is safer than purely human driving, and believes it will become ubiquitous technology around the world.

"Bosch works with all the major car manufacturers across the world," says Srinivasan. "In the UK, we are leading a consortium that benefits from a £5.5 million grant awarded by InnovateUK. The project will see driverless technology trialled in real world conditions on roads in Greenwich, London.

"Automated driving technology in cars will help to prevent accidents, reduce congestion and emissions in cities, offering a more pleasant experience for motorists. However, automated driving is highly complex and requires extensive validation of functions and algorithms involving a large amount of data, to ensure that systems respond to all possible real world-driving situations and are safe."

Safety sells

There is a growing number of people in the industry promoting the idea that fully autonomous cars are safer, but it's unlikely that the general public will agree with them any time soon. UK politicians are aware of this and acknowledged it in a report on the issue.

In a document about autonomous vehicles, the Houses of Parliament says that the Department of Transport "will work to encourage the development and introduction of autonomous vehicles".

However, it notes: "The main policy challenges involve verifying safety and reliability, and creating a legal framework to allow their testing and deployment."

Since that statement, the Government has given the green light to a number of pilot projects, Greenwich being one of them. Others involve Jaguar Land Rover and other companies.

But while the public is sceptical of handing over total control of the steering wheel to the AI systems, increasing numbers of people seem keen on AI assisting them in their driving.

Fully autonomous cars, trucks and other vehicles may not be on the roads for another couple of decades yet, but they're already part of the driving experience for many motorists today. ●

Chapter 6: Alan Norbury

From Art To Science

Interview with Alan Norbury, industrial central technology officer, Siemens UK

German industrial giant Siemens employs approximately 15,000 people at around 30 locations across the UK. Globally, it employs around 350,000 and has annual revenues of more than €75 billion. It is said to be the largest engineering company in Europe. It's also claimed to employ more computer programmers than does Microsoft.

With the European Union referendum looming over the UK, the company's senior executives have made their views clear, with the UK chief executive of the 165-year-old Siemens, Jurgen Maier, saying manufacturing jobs would be at risk if Britain left the EU.

The UK exports almost €150 billion of physical goods to the EU, and it is estimated that around 1 million jobs in the UK are linked to EU trade.

Moreover, some analysts say that for every one job created in manufacturing, several more jobs are created in other sectors, such as information technology, for example, especially now when the internet of things is spreading rapidly across the industrial landscape.

More and more factories and warehouses – the buildings themselves as well as the machinery inside them – are getting hooked up to the IoT and the sector is going through a growth phase. Management consultancy PwC estimates the IoT will lift UK GDP by $300 billion by 2030 at today's levels of investment.

Whether any of this economic activity will be jeopardised if the UK exits the EU is open to speculation for now, but even without the uncertainty the EU referendum brings, the economic environment hasn't exactly been rosy in recent years.

Greater levels of internet connectivity

In an exclusive interview with *Robotics and Automation News,* Alan Norbury, industrial central technology officer at Siemens, explains how C-level executives in the manufacturing sector are dealing with the current economic climate and the trend towards greater levels of internet connectivity in industry, a trend sometimes referred to as Industry 4.0.

Norbury has seen a lot of changes in the global industrial landscape. He started his career as an apprentice in Siemens Congleton, in north-west England, in 1978, moving into the role of technical specialist where he specialised in automation and low voltage products. In 1998 Alan spent time in Germany acting as global account manager for key accounts such as Philip Morris, the famous owner of the Marlboro brand of cigarettes, and ICI, the old archetypal British industrial company whose products included paints and chemicals until it closed down in 2008.

Norbury says: "C-level executives must focus on shifting manufacturing from an art to a science in order to alleviate the strain on manufacturing brought on by the economic climate and current skills shortage.In an interview with EM360, Alan Norbury, industrial central technology officer at Siemens, explains how C-level executives in the manufacturing sector are dealing with the current economic climate and the trend towards greater levels of internet connectivity in industry, a trend sometimes referred to as Industry 4.0.

"The most experienced workers in any industry will have decades of experience in understanding the machinery they work with. However, many of those experienced workers will be retiring in the near future.

"Executives in manufacturing must look to the future, focussing on how we can utilise technology to solve the skills shortage by providing automation or robotics solutions to help ease the transition as experts leave the industry."

Government has the answers

Norbury encourages more consultation with government bodies. He has spent the past two years working closely with the government-funded High Value Catapult Centres, particularly the Manufacturing Technology Centre in Coventry and more recently the AMRC Sheffield.

"Executives in the UK and Ireland need to start better engaging with the likes of government-backed Catapult centres, which focus on business-led collaboration in sectors such as food and beverage, aerospace, pharmaceuticals, manufacturing and oil and gas.

"One of the centres we've worked with is the Manufacturing Technology Centre – the High Value Manufacturing Catapult in Coventry, which helps to bring new concepts into reality, increase focus on innovative products for reconfigurable manufacturing and helps to provide the next generation of skilled workers.

"One of the most important areas executives must consider is hiring the right workers with a grounding in traditional manufacturing processes, with a specific focus on automation technologies and their uses. After all, investment in new technology provides no business benefit if staff are unable to extract maximum value from it."

At the moment, opinion seems divided on whether to or not leaving the EU would be good for the manufacturing sector. Complex regulations may return if Britain were to leave the EU, making it more difficult to do business with many countries in mainland Europe. However, given that so much of the future growth of the global economy apparently depends on China and other emerging economies, some people argue that a so-called "Brexit" would lead to a weaker Sterling and, as a result, make UK exports more affordable to the rest of the world.

2020 vision

Employees at Siemens' Electronics Manufacturing Plant in Erlangen have the freedom to try out innovative ideas and turn them into successful projects.

Siemens has initiated something it calls the Partner 2020 programme, which aims to encourage more investment in technology, particularly robotics and automation technology. It's part of Norbury's brief to develop the latest technology innovations which, while seeking to benefit Siemens UK manufacturing facilities and Siemens industrial customers, could also have a beneficial effect on the UK manufacturing sector as a whole.

Norbury says: "Robotics and automation will play a key role for UK manufacturers looking to increase productivity, however, for those intent on investing, knowing where to turn in the first instance is a key stumbling block. If a company hasn't used automation technology before, the first leap, while often the most important, is also the most daunting.

"Many business owners do not know either who develops the architecture, or where best to procure this technology. Automation within industry isn't only a matter of implementing the technology – the education of manufacturers on the benefits and processes behind automation must become a main focus.

"In order to bridge this information gap, Siemens has recently launched the 'Partner 2020' programme. By utilising expert partner knowledge, manufacturers can break down the initial barrier to automation technology through the deployment of specialist knowledge, insight and guidance, while also allowing access to technologies that can improve growth for manufacturers.

"When developing a brand new factory, manufacturers naturally adopt the most cutting edge technologies, such as robotics and automation. In the UK and Ireland, however, the majority of industries are looking to update existing factories which are in need of modernisation. The UK and Ireland prides itself on running systems on a tight budget, yet the downside is that in the long term we become less productive.

"Copenhagen Business School recently found that if the UK and Ireland were to adopt automation and robotic systems to the same level as certain other countries, for example Germany and Japan, we could increase productivity by up to 22 per cent, and long-term employment by 7 per cent."

Virtual 3D factories

Siemens is such a large company that it could probably dictate the direction in which large parts of the manufacturing industry goes. If, for example, Siemens decided that the way forward for engineering was through the use of virtual reality, it's likely that the largest number of users of VR will be found in the manufacturing sector, and not in computer gaming, as one might expect.

Having said that, Siemens isn't actually the first or only proponent of VR in industry, the technology seems to have found a natural home there anyway. Nonetheless, given its size and influence, what Siemens senior executives like Norbury think of VR and other new technologies is an important indicator of what the future holds for the sector.

Norbury says: "Innovative technologies such as virtual reality, the ability to create fully virtual 3D factory environments to gauge the potential effect of factory changes, will fast become essential to UK manufacturing. A key factor for many organisations looking to upgrade is the perceived risk of operational downtime and unexpected cost; these technologies help to overcome those concerns through parallel innovation.

"Siemens originally utilised this technology to alter factory layouts in order to gain maximum productivity efficiencies, as repositioning physical machines often led to a halt in production, causing a halt in productivity. More recently, we have found undertaking research and development in the virtual world reduces the number of physical prototypes created, with an average of nine prototypes per product reduced to only three. Our long-term goal is to remove the need for physical prototypes, with the entire process run through virtual reality.

"We also expect automation uptake to increase in the consumer market, resulting in faster design of new products and new systems being brought to the market at greater speed. We are already seeing this with food and beverage companies which are allowing customers to specify and adapt ingredients in their food to their specific tastes. Another example is in the customisation and design of bespoke garments, whereby customers are now able to select the design they want and overlay it onto items of clothing.

"Without automation, the scale and speed required with this type of project would be impossible – and cost prohibitive. Automation is shaping manufacturing's future, whether that's unlocking the UK's productivity puzzle or helping to rapidly bring new systems, technology, and products to the market. Used correctly, these technologies have the potential to boost ROI [return on investment] for individual manufacturers across all industries, while also contributing to the country's broader economic output and success." ■

Chapter 7: Scott Mabie

Astronomical Growth

Interview with Scott Mabie, general manager, Americas division, Universal Robots

Industrial robots have always been somewhat captivating to watch because of their precision of movement, but their sheer size and power exudes an air of ominous foreboding, which is why the vast majority have always been caged off, kept away from their human counterparts at factories, where they were mostly to be found.

Large-scale industrial robots perform absolutely essential functions in manufacturing and other industries, but they are potentially very dangerous to humans if they are too close.

That has been the traditional view of robots for a long time. Now, however, things are changing.

A new generation of collaborative robots are being built on a human scale, meaning they have similar arm sizes to humans, and are programmed to be intrinsically safe for humans to use. These new collaborative robots can be, and are being, utilised in situations where they are placed close to humans, often in direct contact or at least within touching, or even hugging, distance. Certainly not caged off.

The appearance, or design, of these new collaborative robots is markedly different from the old robotic beasts of burden seen in factories. These new and advanced robots have been smartened up by their makers, and made to look more presentable for their close encounters with the human kind, with softer colours and more tactile surfaces. Overall, they have become much more aesthetically pleasing and much more ergonomic.

And the emergence of these smarter and cuddlier robots is being embraced in a big way by customers from many different industries.

Exponential growth

Universal Robots, which produces a successful range of collaborative robots for a wide range of industries, has seen exponential growth in its business since its new generation of collaborative robots were launched. In an exclusive interview with RoboticsAndAutomationNews.com, Universal's Scott Mabie, general manager, Americas division, says the company is currently selling more robots in a week than it previously did in a whole year.

Universal Robotics is a Danish company, founded in 2005, by Esben Østergaard, Kasper Støy and Kristian Kassow, who met at the University of Southern Denmark in Odense. It has about 150 employees, and had a turnover of approximately $40 million in 2014. It recently agreed to merge with Teradyne, an industrial automation and testing company, headquartered in the US, which has 3,300 employees and annual turnovers in the region of $1.5 billion. Teradyne is reported to have paid $285 million to acquire Universal.

At the time of the acquisition in May, 2015, Teradyne CEO Mark Jagiela said Universal would add "a powerful, additional growth platform to Teradyne". One of the main reasons for the growth experienced by Universal is its innovate collaborative robots, which Mabie says offer a return on investment – or pay for themselves – in less than 200 days.

Mabie says: "Unlike traditional industrial robots, collaborative robots are lightweight, flexible and can easily be moved and reprogrammed to solve new tasks, meeting the short-run production challenge faced by companies adjusting to ever more advanced processing in smaller batch sizes.

"With traditional robots, the capital costs for the robots themselves account for only 25 to 30 percent of the total system costs. The remaining costs are associated with robot programming, setup, and dedicated, shielded work cells.

"The 'out of box experience' with a collaborative robot is typically less than an hour. That's the time it takes to unpack the robot, mount it, and program the first simple task. Average payback period for UR robots is the fastest in the industry with only 195 days."

Spectacular growth

It would seem inevitable that Universal's spectacular growth was eventually going to lead the company into the upper echelons of the business, where top industrial robot manufacturers such as Yaskawa and ABB reside. But now, with the support of industrial giant Teradyne, Universal's journey to the top may well be a lot shorter.

For Mabie, the merger with Teradyne will introduce entirely new markets to Universal, and enable it to develop more innovative products, and grow as a company, not least in the number of employees it has. In fact, Mabie says his current preoccupation is an "aggressive recruiting" drive to add to the Americas team.

"The combination [of Universal and Teradyne] will boost our ability to innovate and recruit even more and will extend our lead within collaborative robotics and be of benefit to all our end-users and partners.

"At the same time, we are proud to add a brand new line of business to Teradyne. Our operations have been profitable since late 2010.

"Teradyne's world-class engineering and support capabilities and strong financial position will help accelerate the growth of our collaborative robots in new and existing markets, especially in Asia where Teradyne holds a very strong position."

Mabie says Universal will continue to be headquartered in Odense, Denmark, where all research and development, as well as production, is carried out. And even though the company has become part of a giant corporation, Mabie believes it is the products and opportunities that

Universal offers to small and medium-sized businesses that has been the secret of its success so far.

Mabie says: "We've been basically doubling every year, and this year we're expecting to grow 137 percent. Just to put things into perspective, in two or three days this year we'll sell as many robots as we did in all of 2009. That's how much we've been growing.

"We did a little less than 2,000 robots last year. We'll do around 4,000 this year, with revenue close to $100 million. Then next year and the year after, later in 2017, we should hit $200 million. That will be about 9,000 to 10,000 robots.

"I think the reason behind this is that we lowered the whole entry level for utilizing robotics tremendously. Small and medium sized companies that never thought robotics would be within their reach can now automate easily and inexpensively."

That's close enough

The majority of Universal's robots are designed for use in close proximity to humans – they are not caged off. And whereas it can be difficult to assimilate the huge amount of information about the innumerable products most other large robotics manufacturers offer, Universal has simplified its catalogue as much as possible.

"We have over 5,000 robots installed in more than 50 countries worldwide. Around 80 per cent of these operate with no safety guarding.

"Our product portfolio is as simple as our robots. We sell three different collaborative robot arms, all named after their payload in kilos: the UR3, the UR5 and the UR10.

"We launched the UR5 robot in 2008-9, the UR10 was introduced in 2012, followed by the UR3 this spring. Our newest robot, the UR3, is a table-top robot that we're already seeing a high demand for, we expect this to be a game changer especially in the electronics assembly industry."

Like most technology companies, Universal keeps its innovation plans secret. When asked about future products, Mabie would only say that the company has "a comprehensive roadmap of new products and features to be rolled out in coming years", but would not comment on any specific plans.

However, Mabie was more forthcoming on Universal's current

products and markets, giving us an insight into which markets he thinks will grow for the company.

Mabie says: "Globally speaking, our robot sales are roughly 50 per cent in Europe, 25 per cent in North America and 25 per cent in Asia. But we expect the North American and Asian share to start growing now.

"In terms of vertical markets, it really runs the gamut. The machining sector continues to be a big demand driver, but now we're seeing new markets open up and unexpected applications appear.

"Our robots are in applications that receive a 3D laser scan of people's feet and cut out customized flip flops for them, the robots are being tested in agriculture spraying iodine on cow utters before milking to prevent the spread of bacteria, they assemble thermal cups, increasingly handle injection molding machines, and feed CNC machines milling dental crowns and medical devices.

"And as mentioned, we're also seeing a lot of interest from the electronics sector in our new table-top robot – the UR3 – that we launched this spring."

High degrees of accuracy and freedom

Electronics manufacturing requires a high degree of accuracy. So do other manufacturing sectors, although electronics probably has lower tolerances than most.

"With Universal Robots you get a repeatability of one tenth of a mm at the furthest reach of the robot," says Mabie. "You get a tool speed of up to 1 meter per second. You get an ability to move completely free, with 6 degrees of freedom in all directions."

Mabie offers another insight into what is probably a key aspect of the Universal approach, an important ingredient in its recipe for success – and indeed the success of any business or salesman: a thorough understanding of what he is selling.

He says: "The term 'collaborative' not only means that humans can collaborate directly with the robots potentially with no safety guarding between them. The way we see it, the term also addresses the ease of use: a robot is not truly collaborative if it's not easy to work with.

"Our R&D team constantly works on improving what is already the most intuitive and accessible robotics user interface on the market today. You don't need to be online to program a UR robot, it happens through the touch screen on the teach pendant right next to the robot – or by simply grabbing the robot arm to demonstrate desired movement."

The priorities for Universal in designing its robots are:

- easy for anyone to program;
- easy for a customer to install and re-deploy for various applications;
- the robot only requires a 110V outlet; and
- depending on the application the robot can work right next to people without any safety guarding.

And generally speaking, the company's aims are consistent with the age-old purpose of automation. "Universal Robots automate tasks that are dull, dangerous and dirty," says Mabie. "This frees up employees to focus on more rewarding and challenging tasks, which again leads to a better, safer, more rewarding work environment.

"At the same time product quality is improved and an increased competitiveness in the marketplace is achieved."

No need to worry

Mabie is not one of those who believes increasing use of robotics and automation will lead to mass unemployment or anything like that, which is a growing concern for many people in a wide range of sectors in society. Some observers even say that we could see the rise of "new Luddites", taking after the industrial age unemployed people who destroyed the mechanical looms that put them out of work.

"We need to change the notion that robots are here to steal our jobs – all data shows that this is simply not true," says Mabie. "We need to see the robot as a colleague that works right next to us in close collaboration, not a machine that takes over manual labor and gets people fired."

There are differing views on the subject, but most experts seem to be of the opinion that robots will create roughly as many jobs as they eliminate, if indeed they eliminate any at all.

For Universal, and Mabie, robots are a jobs creator. Not only is the company itself hiring more people directly, it is also offering the opportunity for third-party developers to create applications and peripherals for its robots.

A relatively new initiative, the company's URCaps platform opens up Universal robots free of charge to distributors and integrators to build and market end-user solutions, which can be listed and sold through the company's website and other channels.

Universal itself only makes the three robotic arms, no grippers, sensors, vision systems and so forth. All of that comes from partner companies.

Mabie says helping Universal's business partners is of "special interest" to him, and even after 25 years in the industry, he is "having a great time". ●

Chapter 8: Olivier Grenier-Lafond

Tools of the Trade

Interview with Olivier Grenier-Lafond, sales and marketing co-ordinator, Robotiq

Industrial robots have been around for such a long time, and have captured the imagination in such a way, that it would be easy to think that almost all manufacturers have robots installed at their factories.

In fact, the majority of factories in the US do not have industrial robots. And if a country as advanced as the US doesn't have robots all over its manufacturing industry, then no country in the world has a manufacturing sector dominated by industrial robots. Not even Japan.

The vast majority of work done in factories around the world is still done by humans. The Boston Consulting Group estimates the proportion in the US to be around 90 per cent. Meaning, only 10 per cent of manufacturing tasks in US factories are done by industrial robots.

It could be argued that the specific tasks done by industrial robots are the most crucial, but most statistical analyses provide a broad brush picture of any landscape, and the industrial landscape would seem to have far fewer robots than one might imagine.

One of the most important reasons, perhaps the most important reason, for so few robots is that industrial robots have traditionally been very expensive. Purchasing an advanced robotic spot welder would have set you back more than $180,000 in 2005.

A similar machine would have cost $130,000 if you had bought it last year. Still expensive, but significantly cheaper than a few years ago. And prices are set to fall even further – by as much as 20 per cent in the next 10 years, according to analysts. And that probably does not account for the new generation of robots, which are designed using the latest components and techniques.

And, as most people are aware, the thing about technology, especially computing-based technology, namely robotics, is that the more advanced it gets, the cheaper it becomes.

But while the consumer electronics market can offer vast economies of scale to a successful product, niche markets such as industrial robotics move a little more slowly. And that lack of speed may have been a factor in much of the US manufacturing base moving east over the past couple of decades, to countries such as China, which offered enormous human resources even if they couldn't offer advanced robotics and automation technology.

Witnessing the mass migration of manufacturing jobs from West to East, from North America to Asia, were a group of inventive minds in Canada. On seeing more and more factories closing down, having lost out to factories in lower-wage countries, Samuel Bouchard, Vincent Duchaine, and Jean-Philippe Jobin decided that they wanted to do something about it.

They self-funded Robotiq, establishing it in 2008. The company would make robot tooling for agile manufacturing. Their hope was that it would help manufacturers become smarter and more flexible, more adaptable to a changing economy.

While China and Asia still come to mind when one thinks of manufacturing, the US Federal Reserve recently released statistics showing that US manufacturing is making something of a comeback, with modest but consistent growth over the past four years.

For Robotiq, this is good news, as North America has always been their main market, and the region that inspired them to start the company. The company's sales and marketing co-ordinator, Olivier Grenier-Lafond, reiterates the origins of Robotiq, as well as some of its main areas of interest.

Grenier-Lafond says: "Robotiq is a spin-off from Laval University Robotics Lab [in Quebec, Canada], and we just celebrated the company's seventh anniversary. It was founded by Samuel Bouchard, who is the CEO, Jean-Philippe Jobin, CTO, and Vincent Duchaine. We focus on collaborative robots.

"There are two important differences between traditional robots and collaborative robots. The first is that their force limiting, shape and safety features make it possible to use collaborative robots without fencing for some applications.

"The second is that they are easier to program than their traditional counterparts. They can be hand-guided and their user interface is generally more user-friendly. So, they require less specialized knowledge.

"The main reason why this kind of robot is becoming more and more popular is their ease of use. It's easy to learn how to use them, easy to program. This reduces the cost of integration and the cost of training.

"More people in the factory can program these robots. The fact that you don't need fencing around is also interesting because the robot can be repurposed easily within the factory."

This ties in with the Robotiq founders' objective of helping robots to become more adaptable to changing economic circumstances. The company currently sells its products around the world to companies big and small.

Grenier-Lafond says: "We sell through our global network of distributors in around 30 countries to many different customers. From multinational corporations to small and medium enterprises that are just starting to automate their production line. Our robot grippers are also used by many universities and research centers – MIT, CMU, NIST, Fraunhofer, to name a few."

Robotiq has four main product lines, two of which are robotic grippers, one a force torque sensor, and finally a kinetic teaching system which enables robot operators to programme a robot without the need for in-depth programming.

Grenier-Lafond says: "Our most popular product is the 2-Finger 85 Adaptive Robot Gripper. It is designed to be compatible with all major robot manufacturers and it has an almost seamless fit for Universal Robots, the leading brand of collaborative robots. It is mechanically designed to match in size and shape.

"Also we have developed software to work easily with Universal Robots' programming methods. "All Robotiq grippers are designed as programmable robot grippers able to handle most parts in the industrial environment, eliminating the need for changeovers. The grippers can be used in a variety of areas: machine tending, assembly, advanced manufacturing, and research.

"The user has control over the force, speed, and position parameters, allowing for complete control of the gripper."

The Robotiq two- and three-finger grippers of today actually started life as a one-fingered device back on Laval University. It took two short years to go from the one-fingered to the three-fingered version. Overall, however, it has taken many years of hard work to get where the company is today, with its grippers, sensors and kinetic teaching products being used by major companies around the world.

While Robotiq started out thinking about the manufacturing industry, the market that was always going to be its route into other markets was the robotics industry itself. And now, Robotiq grippers are actually very popular with other robotics companies. In fact, the company's 3-Finger Gripper was used by no fewer than eight teams taking part in the Darpa Robotics Challenge, one of the toughest robotic competitions in the world. Two out of the three finalists used Robotiq grippers.

In a way, it's not surprising that they have had an impact on other robotic companies, and have been adopted as one of the "go-to" companies for the more advanced startup companies.

The staff at Robotiq still demonstrate the interest and enthusiasm for robotics that they started with, as evidenced by their daily blogging o the subject, keeping readers abreast of important developments in the news, as well as advances in robotics technologies.

It's certainly appreciated by the many robotics companies that read the posts, as well as the many people who work with robotics, who find the many free Robotiq "how-to" guides useful and informative.

Grenier-Lafond says: "Robotiq writers blog about what drives us as a company and what we consider as the new way of thinking about industrial manufacturing, automation and robotics including: the latest breakthroughs, flexible industrial automation and collaborative robotic.

"The goal is to bring useful content to our partners and to manufacturers looking to automate their production and to support them in doing so. We try to write about different subjects related to robotics in order to reach everyone interested in this field."

The website itself has a clean, open design, which contrasts with many websites belonging to traditional manufacturing companies – even those involved in advanced robotics. Looking at Robotiq's website feels like putting on a spacesuit on your brain, if that makes sense.

It's evidence of the company's collective consciousness, its teamwork. As Grenier-Lafond says, "At Robotiq, the team beats with one heart, this is why teamwork is one of the company's core values." ■

ROBOTICS &
AUTOMATION NEWS

*Market trends and
business perspectives*

Chapter 9: Preview

Next Edition

Interviews

- Dennis Mortensen, CEO, x.ai
- Mike Rigby, head of manufacturing, Barclays Bank UK
- Jalal Bouhdada, founder, Applied Risk
- Ralf Herrtwich, director autonomous technology, Daimler
- Masayuki Morikawa, vice president, Rieti
- Alois Knoll, co-ordinator, Echord
- Erik Walenza-Slabe, CEO, IoT One

www.ingramcontent.com/pod-product-compliance
Lightning Source LLC
Chambersburg PA
CBHW040854180526
45159CB00001B/419